Treating Culturally Diverse Patients?

What You Should Know

DIEULA A CASIMYR MD, MPH

Table of Contents

Acknowledgement

My gratitude goes to Jehovah for my life and everything else He gave me. I want to specially thank my family for their unconditional support.

Preface

My initial interest in writing a book about what Medical Providers should know when treating culturally diverse patients began when I moved to America from Haiti in 2014. I started to work as a

Medical Interpreter and complete a Master's Degree in Public Health as a Foreign Medical Graduate (FMG). Being multilingual, I help those with limited English proficiency communicate with their Medical Providers. I held the position of a Medical Interpreter. I lived for 7 years in Dominican Republic. It helped me understand Hispanic culture . I also worked as a physician in Haiti with Physicians from different Hispanic origin. I began to recognize all the barriers that Haitian Creole natives and Hispanic speaking patients were encountering because of their language and culture.

This book serves as an important tool to help Healthcare Professional to understand culturally diverse populations' perspective. After reading this book your facility will be able to recruit and develop culturally and competent healthcare workforces. When we understand and address how different patients view health culturally health promotion will be better. The burden of disease in our communities will be reduced and culturally acceptable healthcare will become accessible and affordable. Ultimately, this book will enlighten healthcare practitioners so that they can overcome those barriers and continue the path of preventing diseases in patients and managing chronic diseases.

Culture and Health Beliefs

According to Kim Ann Zimmerman (2017), writer and Live Science contributor, "Culture is the characteristics and knowledge of a particular group of people encompassing language, religion, cuisine, social habits, music, and arts." What a group of people does in common and passes to generation to generation is what defines a culture. Every country or group of people is governed by its own culture and that culture influences beliefs about health. Every culture has a different point of view on best health practices.

Per, "Culture and Health", from The Lancet, culture is important in healthcare, as it shapes health beliefs. As the popularity of international travel has increased, it has become more and more necessary that travelers understand how culture affects health beliefs of the population they are entering into. Healthcare providers take care of people coming from all over the world with different health perspectives. Culture should be at the center of Health Care and reassessed in order to serve the varying populations in the United States. Health beliefs are important in the treatment and progress of any disease; hence it is vital that Healthcare Providers build trust with different populations.

Culturally and Linguistically Appropriate Services (CLAS)

In the "Culturally and linguistically appropriate services" (CLAS) it tells us that CLAS is a way to improve the quality of services offered to all people and help to reduce health disparities. CLAS intends to provide healthcare services that are effective and that help people understand one another based on their practices and their primary languages. CLAS promotes the ideas of culturally diverse organizations by offering training programs for all the culturally diverse workforces, support when people are offered services in their languages, and assurance that those providing culturally diverse services are trained to provide the services.

Any organization using CLAS must associate with culturally diverse populations to assess that the cultural and linguistic services rendered are appropriate. Additionally, the organization must have accurate data on the cultural services rendered. Every limited English proficiency person (LEP) in the United States must be aware that language services are available for them in their language, and the individuals providing those language services must be appropriately trained. Providing culturally acceptable services to limited English proficiency people must be the goal of every organization in the country.

Haitian Culture and Health Beliefs

In Cook Ross Inc.'s primer, "Background on Haiti & Haitian Health Culture" (2010), Haitian culture is based on the belief that God can heal everything. Haitians seek natural care with herbs, massages, and over-the-counter medications before they seek medical care. They adhere to medical treatment only if they think the disease is severe enough. Disease can be perceived as a punishment and/or mental disease is very stigmatized in the population. Haitian people are generally friendly and happy. The Primer explains that Haitians' primary health beliefs are based on the balance of different factors such as food, God's will, and personal care. Haitian culture believes that being overweight is a sign of wealth, thus Haitians do not typically exercise for disease prevention.

Haiti's lack of resources, education, knowledge, and money impede on the country's ability to treat the poor for disease. After the natural earthquake that occurred in Haiti in 2010, the country was depleted and left with an even smaller health infrastructure, leaving a lack of health professionals, hospitals, and public health personnel to help its people.

Health Literacy in Haiti

"Compared to native-born American, most Haitian immigrants may have lower levels of health literacy and given their migration to the United States, on average, have lower formal education attainment" (Lubetkin et al, 2015). Lubetkin et al. performed a study to assess health literacy on Haitians Creole speaking patients using a validated health literacy measure: The Brief Health Literacy Measure at the Queens Hospital Center in New York. The age range was 23 years old to 80 years old, with a median age of 52 years old. The Brief Health Literacy Measure has three questions and revealed that most Haitian immigrants have lower health literacy than the rest of the population.

Based on the information cited earlier, little is known about medicine in the Haitian population. Knowledge of medical terminology is limited only to Haitian medical professionals.

Hispanic Culture and Health Beliefs

Dr. Pedro Roma (1982) states that Hispanics share the same Spanish language and Latin, Spain and other Spanish culture countries. However, their ethnic backgrounds are mixed. Most Spanish patients are grateful when Health Care is provided in their native language. Most Hispanics believe that health is a blessing.

Health Literacy in Hispanics

Health literacy in the medical field is the ability to understand health information. Hispanics have suffered from adverse health outcome associated to low health literacy (Jacobson, Hund, & Mas 2016). This is because often Hispanic patients do not understand what the healthcare provider is telling them. For example, a provider informing a patient that he is going to perform a Biopsy may confuse some Hispanic patients if his/her health literacy is low. The Healthcare provider should explain what a Biopsy is if the patients asks. Even when using an interpreter, the interpreter will interpret biopsy in Spanish using the same register as the provider. The interpreter will not explain or describe biopsy to the patient. In my experience working with Hispanic patients they do not ask questions very often. According to two studies, education and age influence health literacy (Mas, Jacobson, & Dong 2014, Soto, Mas, Ji, Fuentes & Tinajero 2015). Healthcare providers should assess a patient's health literacy level at a visit and use simple words to explain procedures with complicated names.

Interpretation Services

The Merriam-Webster dictionary (2017) defines an interpreter as "one who explains." Nowadays with globalization and international travel, people of all languages interact with each other. Some people are proficient in other languages, some can use basic words of other languages to get around on their own in a foreign country, and others can't speak another language at all. "The legal foundation for language access lies in title VI of the 1964 Civil Rights Act, which states: No person in the United States shall, on the ground of race, color or national origin, be excluded from participation in, be denied the benefits of, or be subjected to discrimination under any program or activity receiving Federal financial assistance" (Chen, Youdelman, Brooks, 2007). Interpreters play a vital role in decreasing language barriers between individuals and doctors. Interpreters are professional staff who are proficient in their native language and one or various other target languages. They render the service of interpreting conversations between their client and the person or patient of limited language proficiency. Interpreters can be employees or contractors in different fields such as medical, legal, or social services. Did you know Interpretation is different than translation? Interpretation is done verbally, while translation is done in writing. The New York State department of Labor's website states that on October 6, 2011 Governor Andrew M. Cuomo issued an executive order 26 that requires state agencies to give translation and interpretation services to limited English proficiency people (LEP). Interpretation language services are mandated to be available in all government offices and in private offices where people of all countries go for any kind of medical, social, and legal assistance. "More than 46 million people in the United States do not speak English as their primary language and more than 21 million speak

English less than "very well" (Jacobs et al., 2004).

Cultural Competence

According to the Centers for Disease Control and Prevention (CDC) (2015), "Cultural competence is the integration and transformation of knowledge about individuals and groups of people into specific standards, policies, practices, and attitudes used in appropriate cultural settings to increase the quality of services; thereby producing better outcomes."

The CDC states that all organizations must have structures that allow them to serve and work with diverse cultures, value diversity, and know the community they serve. Cultural competence includes recognizing the complexity of language interpretation. The medical staff of any given federal regulated organization must understand what cultural competence entails in order to serve the population. The concept of cultural competence should not be an option for the Healthcare system; it should be a requirement always. The significance should be properly stressed throughout the health care system because patients' understanding of American Healthcare depends on the Healthcare system being culturally competent.

Insurance Companies in the United States

"Universal health coverage (UHC) means that all people and communities can use the promotive, preventive, curative, rehabilitative and palliative health services they need, of sufficient quality to be effective, while also ensuring that the use of these services does not expose the user to financial hardship" (WHO). Based on that definition of universal health coverage, universal health insurance should include medical, dental, and vision insurance to help everyone in the communities cover for their medical expenses. There're a variety of health insurance companies within the country of the United States, and each one has different prices, different metallic levels that indicate the level of benefits, and different modus operandi. In order to stay compliant with the law the final goal of any health insurance is to cover certain or all medical expenses, depending on the person's age, medical condition, and economic status. health insurance companies in the United States are the link between people and health services. They are used to help people get better and have more economical access to healthcare.

Health Insurance can generally be purchased through the Marketplace, from employers or through the government.

The government objective is to offer universal health insurance. They offer health insurance and tax credits to patients under the poverty level through Medicaid. There are certain conditions to enroll with the government health insurance. The member must comply with requirements such as level of economic status and legal migration status. Medicare is another health insurance offered by the government to people over 65 years old.

Everyone in the country must know how health insurance works, know how to pay, how to enroll, how to unenroll, and understand the benefits that each health insurance covers. When insurance is not paid on time, the individual may lose the benefits and might be responsible for his or her own health costs. Insurance agents help people understand the necessary information when they enroll a member in a health insurance plan. Insurance companies offer interpretation services at no cost for their members who don't speak English.

Health insurance is not popular in developing countries and is a foreign system for immigrants who come from developing countries. The function of insurance companies, deductibles, copayments, modes of payment, and requirements of when to pay are complicated processes for immigrants to the United States. From my own experience with Haitian immigrants and Hispanic patients, many times they receive an invoice from the hospitals and laboratories asking them to be responsible for the medical expenses they incurred. They don't pay the invoice thinking that the invoice was sent by mistake, because they used their insurance card, not knowing that they have a deductible to pay before the insurance company starts taking care of their medical expenses. Immigrants thrive to be on-time and up to date. Even though many immigrants do well in learning the U.S. health insurance system, others still struggle to understand how co-payments and deductibles work, when to pay, and how to pay for their health insurance. They need more information. These immigrants are not familiar with the Healthcare system. Assistance is needed to explain how health insurance works so that they will not lose their health benefits and pay more money for health services.

Insurance companies have the responsibility to invest in their members' well-being. Therefore, they must invest in making sure their members understand the process of the health insurance, especially with culturally diverse populations. Many mistakes are made because of a lack of knowledge. For example after enrollment the first payment must be made without necessarily receive an invoice for the insurance to be active. Because the member did not know that the payment must be made first and did not receive any invoice, the member does not pay until when the member decides to go to the doctor and there is no coverage for that member because of nopayment. During the process of enrollment, Insurance agents must perform due diligence to explain in more detail how and when to pay for one's plan, how deductibles and co-payments work, and what is one's legal migratory status. Only legal immigrants can enroll in health plan and must send proof of their legal migratory status when enrolling and when renewing the health plan to be eligible for benefits.

I have noticed that there is a great need among the elderly Haitian Creole-speaking population for education about the functioning of health insurances in the United States. Oftentimes they don't know how to gain access to interpretation services. Many times, elderly Haitians lose their health benefits, don't pay co-payments on time, lose their tax credits, and must pay high fees for health insurance because they don't renew their immigration documents on time. They don't call to investigate because of lack of knowledge that language services are available to them.

Insurance Companies Responsibilities

Given the fact that Insurance companies exist primarily to cover patients' health care expenses, they also have a responsibility to offer health insurance that benefits their members, even though they are also profit-driven. Because of the Affordable Care Act, insurance companies can't refuse enrollment to members based on their pre-existing conditions. Every member that needs Health care and is searching for health insurance must be able to enroll in a health insurance according to his or her capacity of payment. Insurance companies must invest in the members' benefits and ensure that members with limited English proficiency understand the policies and receive services in their own language.

If people of a certain population or culture constantly have continuous problems with their health insurance payment and other policies, Insurance companies should notice the patterns and incorporate an investigation to discover the problem. The responsibility of health insurance companies is to make sure their members are taking advantage of all the benefits offered and paid for and that they understand the health insurance policies.

Medical Commitment Towards Patients

According to the article, "Medical Oaths and Declarations" (2001), 98% of American medical students take an oath at the beginning of medical school or at graduation. That oath includes everything that focuses on fulfilling the needs of the patients. The patient comes to the practice or hospital for the medical provider; the healthcare provider responsibility is ensuring the patients' necessities are met. Healthcare providers have a moral commitment towards the patients. They are committed to delivering quality health care to everyone, regardless of cultural background, economic status, gender, and language. Once a provider is taking care of a patient, the patient will stay in his care until the patient decides he or she no longer needs the services, or until the patient is cured. The Healthcare Provider's commitment is to medically treat the patient without prejudices and ultimately to keep all the conversation between the provider and the patient confidential. The Healthcare team exists to promote health and life. The medical provider can't promote death. The Healthcare team has the responsibility to ensure that the patients understand their care, and that they receive care in their language. It is important for Healthcare teams to understand the health culture of the patients they are taking care of because that promotes health and ensures that patients can follow the treatment and get treated.

Medical doctors must remember and vow to keep confidential information about each patient secure and private according to HIPAA. The patient-provider relationship must be kept private all the time. When a health provider uses a language service to communicate with a patient, the patient is more responsive because he or she feels that the physician understands him or her.

As Haitian I know that Haitian Creole-speaking patients are particularly sensitive to their language. When someone who speaks their language talks to them, they feel at ease and lose the fear of talking and telling the medical provider all the symptoms they are feeling. From my personal experience, Haitians immigrants always migrate in places where there are other Haitians because they feel understood. When they have an interpreter present to help with questions, they feel reassured. However, most American medical providers do not understand Haitian and Hispanic health culture, therefore, many providers do not know exactly what their patients are feeling or thinking. In order to ensure that good, quality care is delivered, medical providers must make sure that their culturally diverse patients feel understood and that they have said everything they wanted to say during the visit with the doctor.

Prevention and Care Continuation

The World Health Organization (WHO) defines disease prevention as "understood as specific, population-based and also individual- based intervention for primary and secondary (early detection) prevention, aiming to minimize the burden of diseases and associated risk factors". Preventive doctor's visits are important to avoid illnesses and diseases. Simply by washing hands, many illnesses can be prevented. Public health professionals have a responsibility to educate the population about disease prevention. "Health promotion is the process of empowering people to increase control over their health and its determinants through health literacy efforts and multisectoral action to increase healthy behavior" (WHO). Patients need to understand the importance of prevention to practice it and to reduce the burden of diseases. Stephen Schimpff medical doctor and author of, *The Future of Medicine states* "Prevention is the key to both better health and lower health-care costs over the long haul. This is where the nations and each of us as individuals needs to put energy and resources". Prevention dramatically reduces cost, hospitalization stays, and doctor's visits for illness.

The purpose of medical treatment is to cure acute illness and, in the case of chronic diseases, give treatment to palliate the symptoms, prevent additional disease, and give patients a better quality of life. If a patient does not adhere to treatment, pick up medications at the pharmacy, or follow their provider's instructions, the consults will not give any positive results.

Depending on how patients view healthcare and its importance, it is possible that they will not follow up with their care in picking up medications, consistently scheduling a follow-up appointment, and taking the medications as the doctor prescribed them. This frequent scenario is particularly true in the case of

Haitian and Hispanics. From personal experience with limited English Proficiency patients, there is a lack in care continuation. To illustrate after the doctor's visit, patients often ask for repetition of the doctor's instructions. Care continuation is an important key to avoid re-hospitalization or complications of acute and chronic illnesses. With proper care continuation, Haitian Creole-speaking patients and Hispanics will know what to do after each visit, how to take their medications, how to protect themselves, and when to come back for the follow-up appointment. The care continuation team, such as Care Managers and Care Coordinators, must stress to their patients the importance of following up, especially those with culturally diverse backgrounds.

When treating culturally diverse patients, the Healthcare team shouldn't assume that their patients know everything about healthcare, which is why cultural competency is important in healthcare. A culturally competent healthcare team must know how to approach a patient and what areas of care to explain so that the patient can fully understand the process of care. In sum, when the patient and the healthcare team are on the same page, care continuation is more effective and give good results.

Cultural Competency

Cultural competency is very important in health care to reduce the burden of diseases and to serve a culturally diverse population. In the case of Haitian Creole-speaking patients and Hispanic, the Healthcare team must be knowledgeable of the educational and cultural backgrounds. This will help avoid mismanagement and confusions about terminologies, preferred language, disease denominations, use of interpreter services and other barriers. Being Haitian, myself helps me to understand that when Haitian Creole-speaking patients visit a hospital or a clinic, they expect to get treatment, moreover, if they are limited English proficiency patients, they also come with a fear of the language and of the unknowns of an illness. If they encounter language services, they are relieved because they can rely on an Interpreter. Below are some of the main barriers Haitian Creole-speaking patients and Hispanics face in the United States health care system identified from own experience while interacting with Limited English Proficiency patients.

Medical Terminologies

"The terminology of a subject is the set of special words and expressions used in connection with it" (English dictionary). The Healthcare system has its own terminologies used to reference procedures, illnesses, diseases. Healthcare interpreters must maintain the same register as the healthcare provider they are working alongside, because they are not allowed to explain to the patient what the healthcare provider might mean, nor can they ask the patient if he or she understands. The interpreter simply serves as a conduit for the message, and a tool to decrease language barriers between healthcare professionals and patients. Therefore, the interpreter is not part of the conversation. If a Healthcare provider's language is not clear, the patient either guesses what the provider means or says "No" because he or she does not know the medical terminology. Few patients ask for an explanation of the terminology. As an example, if a Healthcare provider asks a patient, "Did you have a bowel movement today?" the interpreter will ask the same in Haitian Creole or Spanish. However, depending on his or her educational background, the patient might not know what that means. Per experience, some patients might ask what that is, but some might not ask and remain with the doubt and guess at the answer. Explaining terminology is very important for the patients to understand and give the correct answer so that they can receive appropriate treatment and care.

One term that is used often in healthcare and is confusing to Haitian and Hispanic patients is, "Transfusion." The medical interpreter interprets "Transfusion" in Haitian Creole or Spanish, but the patient might not know what that means. It is important to know if a patient has had a previous transfusion to assess previous allergies to blood transfusions and to get consent to another transfusion. The provider must make sure the patient knows what each medical term means. For limited English proficiency patients, it is important that providers define terminologies because the interpreter will not define the terminology. Some of the most common terminologies that can easily be misunderstood are endoscopy, colonoscopy, CT scan, MRI, PPD, MRSA, living will, and advanced directive. The healthcare team is responsible for ensuring that their patients understand the terms they use, know what each term may entail, and understand why the patient may need a certain procedure.

Medical terminologies are one of the barriers that Haitians and Hispanics experience in American Healthcare.

French and Haitian Creole

According to Gracia et al. state that "French and Creole are the official language of Haiti. French is the principal written and administratively authorized language. It is the preferred medium of instruction in most of the schools. Most (approximately 90%) of Haiti's 10.3 million people, however, use Creole as their primary language". From growing up in Haiti, I can attest that not everyone speaks French in Haiti; even some people who do go to school do not speak French, as this largely depends on the school's quality and of the individual's background. Some Creole speakers may understand French, but can't reply in French. Very often, someone who speaks French in Haiti will also speak English. Depending on the educational and social background of an Haitian individual, he or she will speak French or Haitian Creole, or French and Haitian

Creole, and may additionally be fluent in one or two other foreign languages. However, there is no such language as French- Creole in Haiti. The combination is not a language.

Healthcare professionals must make sure that the language spoken and understood by the patient is the same as that of the interpreter. Addressing a medical patient who speaks Haitian Creole in French will create an even bigger gap between the healthcare team and the patient, even with an interpreter. It is better if the healthcare team asks each individual Haitian patient what his or her preferred language is; this way, the patient feels more confident to say what his or her preferred language is without feeling pressure or an obligation to try to understand a language he or she is not fluent in or partially understand.

Patients will receive a high quality of care and can receive the medical help they need if they are speaking and listening to their native tongue. Establishing the difference between patient who speak Haitian Creole and patients who speak French is another barrier that Haitian face in healthcare.

Blood Pressure

Uwe Diegel specialist in various types of medical diagnostics states, "It is interesting to note that in most French speaking countries blood pressure is expressed in cmhg mercury, which means that the doctor will express your blood pressure as 12/8(twelve over 8 instead of 120/80" This is the case in Haiti. A culturally competent healthcare team must be aware of that difference to treat Haitian Creole-speaking patients properly and be able to know what the equivalence is in English. A Haitian Creole-speaking patient might not be aware of that difference and think that the reading is the same everywhere.

I know of a case that happened during a normal visit with a provider, a Haitian Creole-speaking patient said that her blood pressure normally runs to 14/7 and had the provider confused. This is another barrier faced by Haitian immigrants.

Weeks of Gestation

According to the article "How to calculate your pregnancy due date in France", doctors in France calculate women due date counting 2 weeks after the first day of period and count 9 months. Personally, I know that Haitians count weeks of gestation in months. They count each month of pregnancy until the nine months, when they are at term. In the United States, however, the gestation period is counted more frequently in weeks. Thus, when asked for the weeks of gestation, a Haitian patient in America would be confused. However, if the provider knows the cultural differences of his patient, the provider could clarify what he means.

Pain Scale

"Haitians can use numerical scales for symptoms if the scales are explained" (Colin). Not everyone outside of the medical field is used to the 1-10 pain scale used in American hospitals. When a Haitian Creole-speaking patient is asked to identify the degree of pain that they feel on a scale from 1-10, it is likely that the patient does not know what that means. Because of that, the patient might simply reply that they're in pain. Healthcare professionals are often required to document a specific number in their charts. To Haitians, pain is simply pain; there is no pain scale in Haitian culture.

The Healthcare professional team must help the patient understand how to classify pain to get an accurate answer. This is another barrier that Haitian immigrants face in healthcare.

Care Management

"Haitians seek medical care when it becomes clear that an illness requires attention" (Collin). Care management is another new aspect for immigrants; these diverse patients don't usually understand what it entails, why there is a Care Management team, and what they can do with their care manager.

When offering services to culturally diverse populations, proper emphasis must be placed on the availability of language services in order to eliminate the language barrier. A step-by-step plan must be followed to help the Haitian Creolespeaking patient get used to the service and follow the plan accordingly.

Elderly Patients

Relating my experience, when elderly Haitian Creole and Hispanics patients are hospitalized, the healthcare team use an interpreter service to communicate with them, both in person and over the phone. Sometimes when utilizing over the phone interpreters phone conversations can leave both the patient and the healthcare team frustrated. When the elderly Limited English Proficiency patients select to use an over-thephone interpreter, they often can't hear well because of hearing problems. Therefore, it is usually better to use a face-to-face interpreter with elderly patients to avoid confusion and a waste of time.

Discharge Instructions and Medications

Very often, with several patients that I have interacted with, when Haitian Creole and Hispanic patients are getting discharged using an interpreter, they must keep track of a lot of information at once. This may cause them to lose information once they return home. They may not follow the treatment properly. To avoid readmission, Healthcare providers should follow up with these patients after discharge to assure that the patients are following the instructions properly.

Follow-up Appointment with Analysis

Every issue comes back to the language barrier that is difficult to bridge in medical care. From experience many limited English proficiency patients do not know that they have interpreters available to them, and Healthcare providers often do not hire interpreters or let patients know that interpreter services are available to them. To illustrate this problem, I have a story: A limited English proficiency patient was asked to return to a laboratory in three months to complete an analysis, then asked to return for his follow up appointment, when he returned to the doctor, he came without the analysis. This occurred because he did not know how to go to the laboratory, what to say, or how to ask for an interpreter. Consequently, his follow-up care was delayed; the provider sent the patient for the analysis again, wasting both the provider's and the patient's time. Had the services been explained clearly at the beginning of the patient's care, the problem could have been avoided.

These barriers in the medical area are just a few problems that Haitian Creole-speaking patients and Hispanics encounter based on my experience. Cultural competency is vital in medicine to deliver high quality care especially to Haitian Creole-speaking patients and Hispanics. Ultimately, Healthcare professionals must pay close attention to culturally diverse patients and strive to understand their culture. They should ensure their patients understand the care being delivered to them and that the care is culturally acceptable to them.

Barriers to Insurance Services For Haitian Creole-speaking Members and Hispanics

It is extremely important that Haitian Creolespeaking patients and Hispanic understand their insurance company and plan, that they know how to use their ID numbers, that they understand what benefits they are paying for, and that they become familiar with the documents to enroll and renew their plan. At the time of enrollment, insurance agents should emphasize the basic elements necessary to an insurance plan for the culturally diverse population. If this is done properly, limited English proficiency people can be involved in their medical insurance process and avoid many barriers they often encounter. Medical insurance staff must be culturally sensitive and culturally competent, always considering that culturally diverse populations may have different views about medical insurance.

Verification Process

When a patient decides to call his insurance company, they must go through a verification process. Because of HIPAA and lack of knowledge about the insurance process, I found out based from experience that culturally diverse patients are often reluctant to give their information, alleging that they are already a member and that the insurance company has their information in the system. They may not understand that the insurance agent is required to verify everyone who calls. When the agent tells them that the call is being recorded or monitored, they often do not understand why. This problem can be avoided if it is explained to the member during enrollment that it is standard procedure to verify each member's identity when he or she calls.

Spelling of Name

Jessie Colin states, that "80% of Haitians neither read nor write" This is also true for other cultures. It is an unfortunate reality. This causes problems for illiterate insurance members because they can't spell their names for the verification process. There should be a provision for those members when they call over the phone, so that they would be able to verify their identity.

Verbal Authorization Process

Depending on the patients' health status, I have experienced that they can't have a conversation, and they need a family member or a guardian to speak on their behalf. Not everyone knows the HIPAA laws that protect patients' and members' privacy. Culturally, for Haitians and Hispanics a close relative such as a spouse, child, or parent, can talk on behalf of his or her relative who is old, feeling ill, or mentally handicapped without prior authorization from the member.

Thus, when those people are told that they require prior authorization from the party in question to release personal information, they get frustrated because they do not understand why, seeing as they are close family members. Those confusions can be avoided if the insurance company clearly and consistently explain these requirements beforehand when the patient enrolls in the insurance plan.

Preventive Health Assessment/Yearly Call

I know from experience that some immigrants struggle with the preventive health assessment or yearly call. The preventive health assessment is used to assess the health of the members and help them prevent complications. However, many members are suspicious; they do not want strangers in their home to assess them, so they decline the visit. Consequently, these members do not take advantage of the benefits that the medical insurance offers. If the Care Manager or the enrollment agent explains this assessment, the members will be more receptive to the visit. Often, the medical insurance companies send a letter explaining and announcing the visit. However, these members often do not read the letter because they do not speak English and there is no one at home to translate the letter for them.

Culturally Competent and Knowledgeable Staff

It is important for the staff to be culturally competent and knowledgeable. When the staff is culturally competent, they can be sensitive to the needs of the culturally diverse people they are serving. Additionally, when they know medical terminologies they are better equipped to help the members without relying on the member to spell everything for them. Cultural competency is important for medical insurance staff because they serve a large population of diverse cultures.

These barriers with the medical insurance companies highlight the need for a culturally competent staff to understand the needs of their members and be explicit in conversation. They should use all the resources available to make limited English proficiency people understand how the insurance system works.

Overcoming the Medical and Insurance Barriers

When the healthcare team understands how important it is to make healthcare culturally acceptable for culturally diverse populations, healthcare promotion, care continuation, care management, and treatment will be easier for patients and providers. The burden of disease will be reduced, the healthcare gap for minorities will be reduced, and culturally diverse patients will feel involved in their care. Culturally diverse patients will be motivated to work with their healthcare teams to get better, to follow up with their appointments, to pick up their medications, and they will know all the services available to them. They will know to ask for an interpreter to have access to medical and insurance services. A culturally competent healthcare team is good for the culturally diverse population they serve.

Unnecessary insurance costs can be avoided and reduced greatly when the members know how to use their services and when they know what their benefits entail. Insurance company staff should have a better approach with culturally diverse members to assure a deeper understanding of insurance companies.

Suggestions and Limitations

Eliminating those language barriers and others that may come up will decrease the gap in healthcare for minorities and culturally diverse patients.

Here are some suggestions that I recommend to overcome the barriers that culturally diverse populations face. First, communication is the key to overcome the barriers that those patients encounter. The rights and expectations of every culturally diverse patient should be clarified for the entire healthcare team. Hence, certified medical interpreters should be used whenever a Limited English Proficiency, culturally diverse patient is receiving medical care. Do not use a family member because it might be faster or a bilingual medical assistant from the practice. Both family member and medical assistant are neither certified medical interpreters nor versed in medical terminologies. Second, there should be a commitment to serve culturally diverse patients in their language and to ensure that they accept and understand the care that is being delivered to them. Assuring culturally acceptable healthcare services takes time, work, and training of the whole healthcare team about cultural competence and acceptance of diversity. Third, the healthcare team must be aware of the barriers in healthcare, leaving behind any assumptions, misunderstanding, prejudice, and fears. Fourth, insurance companies can eliminate the barriers for the culturally diverse populations by explaining in more cultural details all of the unknown medical, billing, and insurance terms to their members, not assuming that the members understand the terms. Detail every aspect of insurance enrollment for the members; be sensitive to cultural health views and differences; be culturally competent; and promote language services to all culturally diverse populations, ensuring that they know they can communicate through an interpreter every time they call for a service. Fifth, every year cultural consultants should assess and evaluate the progress and services that hospitals and insurance companies offer to the culturally diverse population. The limitations for those barriers are that there are simply not enough training forums for cultural competencies, not enough emphasis for care continuation, and not enough diversity in the workplace. To reduce the gap between the healthcare team members and culturally diverse populations, the limitations and barriers must be overcome because the concept of fear of change in the workplace is not productive at all.

References

CDC. 2015. "Cultural Competence." (NPIN) National Prevention Information Network.

https://npin.cdc.gov/pages/culturalcompetence

Colin, J. M. "Cultural and clinical care for Haitian." https://www.in.gov/.../Haiti_Cultural_and_Cl inaical_ Care_Presentation_Read-Only.pdf

Cook Ross Inc. 2010. "Background on Haiti & Haitian Health Culture." https://www.in.gov/isdh/files/ cultural_prim

er_on_Haiti.pdf

Definition of terminology. https://www. collinsdictionary.com/dictionar

y/english/terminology

Diegel, U. The white book on hypertension (1st ed.).www.medactiv.com/downloads/MedAct iv-Blood-Pressure-FAQ-pdf

Expat Financial. N.d. "Haiti Insurance." expatfinancial.com/regions/Caribbean/Haitiinsurance/

Haiti Outreach Ministries Resources. 2014.

"Haitian Health Care Beliefs and Voodoo

(Voudou)."www.haitiom.org/wpcontent/
uploads/2014/10/Health-carebeliefs-and-Voodoo.pdf

Jacobs, E. A., Shepard, D. S., Suava, J.A., S, E. 2004.
"Overcoming language barriers in healthcare: costs
and benefits of interpreter services." American Journal
of Public Health,95(5):866-

869. https://www.ncbi.nlm.nih.gov>NCBI

>Literature > PubMed Central (PMC)

Jacobson, H. E., Hund, L. Mas, F. S. 2016.

Predictors of English Health Literacy among U.S.
Hispanic Immigrants: The importance of language,
bilingualism and sociolinguistic environment. Literacy
and numeracy studies

24(1): 43-64

"The Language access and assistance." https://
labor.ny.gov> Div. of Immigrants Policies and Affairs

Le Bras, C. 2011. "How to calculate your pregnancy
due date in France." French Mama Pregnancy and

Parenting in France. frenchmama.com/2011/07/25/
how-tocalculate-your-pregnancy-due-date-in-France

Lubetkin, E. L., Zabor, E. C., Isaac, K., Brennessel,

D., Kemenv, M. M., Hay, J. L. 2015. "Health

Literacy, Information Seeking and Trust in

Information in Haitians." American Journal of
Health Behavior 39(3): 441-450. doi:

10.5993/AJHB.39.3.16

Merriam-Webster. "Interpreter." https://
www.merriam-

webster.com/dictionary/interpreter

Nadia, J. N., Brown, C. M., Sampson, D. M., Ejike-
"Haiti."King, L

https://wwwnc.cdc.gov/travel/yellowbook/2

018/select-destination/haiti

Napier, D. A., Ancarno, C., Butler, B., Calabrese, J.,
Charter, A., Chartterjee H., and Woolf, A. 2014. "Culture
and Health." The Lancet 384(9954):1607-1639.doi:

http://dx.doi.org/10.1016/S0140-

6736(14)61603-2

Poma, A. P, (1983). Hispanics Cultural Influences on Medical Practice. Journal of the National Medical Association 75(10): 941-946.

https://www.ncbi.nlm.nih.gov>pmc> articles

>PMC2561612

Sritharan, K., Russel, G., Fritz, Z., Wong, D., Rollin,

M., Dunning, J., Morgan, P., ... and Sheehan, C. 2001. "Medical Oaths and Declarations." British Medical Journal, 323 (7327), 14401441.

Pubmedcentralcanada.ca/pmcc/articles/ PM C/1221898/

U.S. Department of Health & Human Services. N.d. "National Culturally and Linguistically appropriate Services Standards." .

https://www.thinkculturalhealth.hhs.gov/cla s/standards.

World Health Organization. 2017. "What is Universal Coverage?" www.who.int/health_financing/ universal_co verage_definition/en/

Zimmermann, Kim Ann. 2017.

"What is Culture? Definition of Culture." Live Science. https://www.livescience.com/21478what-is-culture-definition-of-culture.html.

About The Author

Dieula Casimyr is a proud Haitian born Medical Doctor and Master of Public Health who began working as a Medical Interpreter in 2015. She earned her Medical degree at the "Universidad Católica Tecnológica Del Cibao in Dominican Republic and her Masters of Public Health at the University of West Florida in Pensacola. Being Multilingual, she relates with culturally diverse populations and endeavors to significantly reduce the language barriers between culturally diverse patients and their providers. As a volunteer in CapraCare, she was part of the leadership for the Infection and Prevention committee, implemented in Haiti and presented a poster about intimate partner violence

(IPV) and unintended pregnancy (UP) in minorities in United States at the American Public Health association (APHA)'s 2017 Annual Meeting and Expo in Atlanta.

Dieula Casimyr continues to work for minorities and culturally diverse patients as a Public Health professional and cultural consultant with the final goal of promoting health equity and decrease health disparities.

Becoming an Author is now added to her long list of accomplishments in her career. Dieula first book titled, **Treating Culturally Diverse Patients? What You Should Know** is available for Pre-order now!

Contact

Dr. Dieula Casimyr MD, MPH Training and Conferences:

Casimyrcompetency.com

Email: dcasimyr@gmail.com